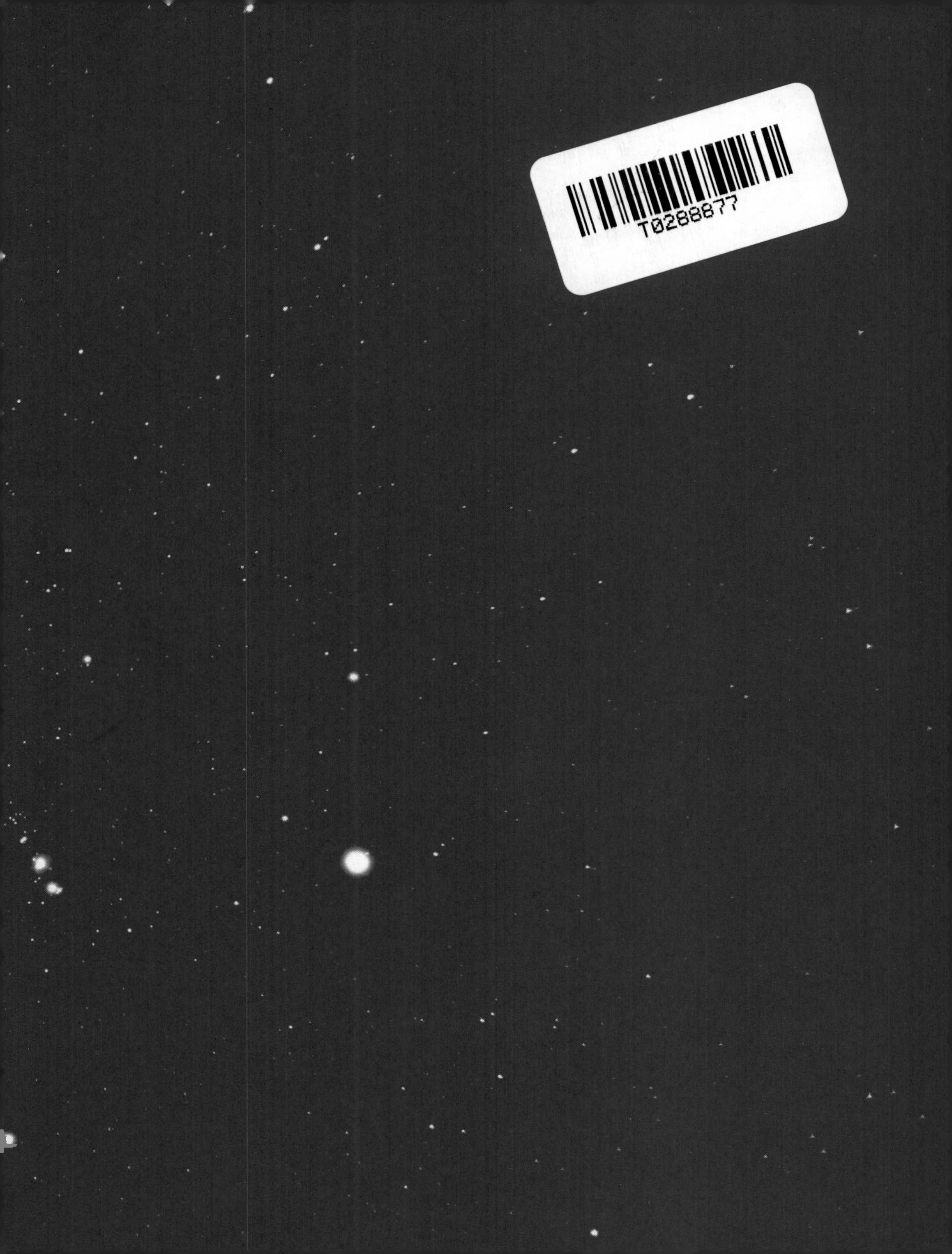

CELESTIAL

CELESTIAL

A LOVE POEM

Abhay K.

MAPIN
PUBLISHING

First published in India in 2024 by
Mapin Publishing Pvt Ltd.
706 Kaivanna, Panchvati, Ellisbridge
Ahmedabad 380006 INDIA
T: +91 79 40 228 228 • E: mapin@mapinpub.com
www.mapinpub.com

ISBN: 978-93-94501-15-7

Design: Rakesh Manger / Mapin Design Studio
Copyediting & Proofreading: Mapin Editorial
Production: Mapin Design Studio
Printed in India

Acknowledgements: I would like to thank Garima Garg for encouraging me to
publish this poem as an illustrated book. I am grateful to Chaitali Pandya for
reading the poem carefully and offering her valuable suggestions. My gratitude
to Bipin Shah for offering his valuable suggestions and publishing the book.

Poet's Note

The thought of writing *Celestial* came to me during my stay in the Indian Ocean island of Madagascar from 2019 to 2022.

Antananarivo skies are clear and free from light pollution during most part of the year. I started gazing at the night sky more attentively and was duly rewarded with sightings of the planets in our solar system. I could see some prominent star constellations such as Orion, Canis Major, Taurus, Scorpius, Centaurus, Vela, Carina, Sagittarius, Libra, and bright stars such as Sirius, Canopus, Betelguese, Rigel, Aldebaran, Antares with the naked eye. This also set me on the path of writing anthems on the planets in our solar system, which were published on 21st December 2020, on the day of the Great Conjunction of Jupiter and Saturn.

Fascinated by the stories and mythologies associated with constellations, I started reading about them in depth. Once, quite late into the night I woke up from a deep sleep and looked through my window towards the sky and a poem started taking shape in my mind on seeing the constellation of Scorpius shining bright in the eastern sky.

During my university days, I had read a few lines of the poem *Locksley Hall* by Lord Alfred Tennyson:

> *For I dipt into the future, far as human eye could see,*
> *Saw the vision of the world, and all the wonder that would be;* 6

Till the war-drum throbbed no longer, and the battle-flags were furl'd
In the Parliament of man, the Federation of the world. 64

There the common sense of most shall hold a fretful realm in awe,
And the kindly earth shall slumber, lapt in universal law. 65

However, it was in Antananarivo that I got the opportunity to read the complete *Locksley Hall*, which was inspired by Sir William Jones's prose translation of *Mu'allaqat*—a group of seven long Arabic poems. The poem consists of 97 rhyming couplets and makes reference to the constellation of Orion and asterism Pleiades:

Many a night from yonder ivied casement, ere I went to rest,
Did I look on great Orion sloping slowly to the West. 4

Many a night I saw the Pleiads, rising thro' the mellow shade,
Glitter like a swarm of fire-flies tangled in a silver braid. 5

Inspired by this poem, I set out to write *Celestial*, consisting of 100 rhyming couplets that could take the readers on a cosmic journey to all the star constellations visible from Earth, some bright, some faint, some visible only in the northern hemisphere, and some visible only in the southern hemisphere.

The International Astronomical Union (IAU) recognizes 88 star constellations covering the entire sky in both the hemispheres. A constellation is defined as "an area on the celestial sphere in which a group of visible stars forms a perceived outline or pattern, typically representing an animal, mythological subject, or inanimate object."

Naming of constellations predates history. Each culture had its own set of star constellations and their unique names until the IAU officially recognized 88 constellations in 1928. Out of these, 48 constellations are traditional Greek constellations visible from the northern hemisphere, listed in *Phaenomena* by Aratus (310–240 BCE) and *Almagest* by Ptolemy (100–170 CE). In fact, *Phaenomena* is a poem written in hexameter, first half of which describes the constellations and other celestial phenomena, while *Almagest* is a mathematical treatise on the apparent motion of the stars and planets. Later, 40 more constellations were added between the 15th and 18th centuries when explorers started travelling to the southern hemisphere.

Most of the constellations have Latin names and are named after classical Greek legends though they had been detected much earlier. For example, Orion, one of the most visible and prominent constellations, has been seen as a hunter since Babylonian times. The constellation of Scorpius was known to the ancient Egyptians and is addressed as Ip in Egyptian hieroglyphs. In Indian mythology, Orion is the warrior Skanda—the six-headed son of Shiva, while the seven central stars of Ursa Major that form the kettle are referred to as Saptarshi meaning seven (*sapta*) sages (*rishis*).

In Indian astronomy, a *nakshatra* is one of the 27 sectors along the ecliptic. There are 27 *nakshatras* listed in *Atharva Veda* and their names are associated with the prominent stars or asterisms. For instance, Krittika corresponds with the Pleiades, Rohini with Aldebaran, Ardra with Betelgeuse, Punarvasu with Castor and Pollux, Magha with Regulus, Uttara Phalguni with Denebola, Chitra with Spica, Svati with Arcturus and Satabhishak with Sadachbia.

Of late, there is an emerging global trend of astro-tourism, which means people go stargazing to ideal locations that are free from light pollution and have darker skies. Bortle Scale measures the darkness of the skies at any given location on a scale of 1 to 9. A scale of 1 means an ideal dark sky location with most of the constellations visible with the naked eye, while a scale of 9 means heavy light pollution as in the large cities.

International Dark Sky Association's mission is "to preserve and protect the night time environment and our heritage of dark skies through quality outdoor lighting." It does this by certifying certain places as International Dark Skies places. There are now more than 190 such places[†] in the world. India too has a number of dark skies places ideal for stargazing, including Pangong Lake in Ladakh, Nubra Valley in Leh, Benital in Chamoli district of Uttarakhand, Rann of Kutch in Gujarat, Neil Island in Andaman and Nicobar, among others, and the interest in astro-tourism is growing.

Darkness has its own beauty. We should preserve dark skies so that we can revel in the light of the universe.

I hope this poem inspires you to relish the beauty of starlight, and to preserve and protect our heritage of dark skies.

Abhay K.
Antananarivo, Madagascar

[†]https://www.darksky.org/our-work/conservation/idsp/

Note on Illustrations

This book contains illustrations of 48 constellations from the book *Ṣuwar al-kawākib* or *The Book of the Fixed Stars* by the renowned tenth-century Persian astronomer 'Abd al-Raḥmān al-Ṣūfī (903– 986 CE, Rayy, south-east of Tehran), known in the West as Azophi. The book was originally written in old Arabic without commas or full stops and contained illustrations, a detailed catalogue of stars, their magnitudes, colours and coordinates. Al-Ṣūfī's original manuscript has been lost, but

the Bodleian Library, Oxford, has an early copy reputedly made by his son Ibn al-Ṣūfī 24 years after his father's death, although it is considered a much later copy by a different author. This manuscript is known as Marsh 144 at the Bodleian Library. Incidently, Ibn al-Ṣūfī also wrote a poem about the stars to make his father's work popular. The Bibliothèque nationale de France has a manuscript of *Ṣuwar al-kawākib* that was prepared for the mathematician and astronomer Ulugh Beg of Samarkand, grandson of Tamerlane, in 1436.

Al-Ṣūfī's book was based on Ptolemy's classic work *Almagest*. He updated Ptolemy's measurements of longitudes of the stars by adding 12 degrees and 42 minutes, to accommodate precession.

Al-Ṣūfī produced dual illustrations for each of 48 constellations, one portrayed on the celestial globe while another as viewed directly in the night sky, the main attraction of the book for a layman, something missing in the original *Almagest* of Ptolemy. In each illustration, stars that Ptolemy catalogued in *Almagest* are shown in golden colour on the constellation diagram, with labeling in black. Stars observed by Al-Ṣūfī are depicted in red. Al-Ṣūfī inserted more than 40 stars on the chart from his own observations besides charting Ptolemy's stars from *Almagest*. Here is a list[1] of constellations illustrated by Al-Ṣūfī along with their Arabic names which are almost literal description of their Greek names:

The twenty-one northern constellations:

1. Ursa Minor – *ad-dubb al-aṣġar*, "the Smaller Bear"

2. Ursa Major – *ad-dubb al-akbar*, "the Greater Bear"

3. Draco – *at-tinnīn*, "the Serpent"

4. Cepheus – *al-multahib*, "the Burning One"

5. Boötes – *al-ʿawwāʾ*, "the Howling Dog"

6. Corona Borealis – *al-iklīl aš-šimālī*, "the Northern Crown"

7. Hercules – *al-jāṯī (ʿalā rukbatihī)* "the One Kneeling (on his Knees)" (from the Greek *ho en gonasin*, "the One on his Knees")

8. Lyra – *an-nasr al-wāqiʿ*, "the Falling Eagle" (from the medieval Latin name Aquila Cadens)

9. Cygnus – *ad-dajāja*, "the Hen" (from the Greek *ornis*)

10. Cassiopeia – *ḏāt al-kursī*, "the Woman on the Chair"

11. Perseus – *ḥāmil raʾs al-ġūl*, "the Carrier of the Monster's Head" (meaning Medusa's head)

---

1. Classical Astrologer. https://classicalastrologer.com/the-48-ptolemaeic-constellations-found-in-the-almagest-al-majis%E1%B9%ADi/.

12. Auriga – *mumsik al-ʿinān*, "the One Holding the Reins"

13. Ophiuchus – *al-ḥawwā ḥāmil al-ḥayya*, "the Snake Charmer Carrying the Snake"

14. Serpens – *ḥayyat al-ḥawwā*, "the Snake of the Snake Charmer" (from the Greek ophis ophioukhou)

15. Sagitta – *uwīsṭus* (from Greek oïstos); *as-sahm*, "the Arrow"; *an-nawl*, "the Loom" (from an erroneous reading of *oïstos*, "arrow", as *histos*, "loom"); *al-ġūl*, "the Monster" (an erroneous reading of *an-nawl*)

16. Aquila – *an-nasr aṭ-ṭāʾir*, "the Flying Eagle"

17. Delphinus – *ad-dulfīn*, "the Dolphin"

18. Equuleus – *raʾs al-faras*, "the Head of the Horse" (from the Greek *hippou protomē*, "the Front of the Horse"); *al-faras al-awwal*, "the First Horse"

19. Pegasus – *al-faras al-mujannaḥ*, "the Winged Horse"; *al-faras al-aʿzam* "the Great Horse"; *al-faras aṭ-ṭānī* "the Second Horse" (from the Greek *hippos*, "Horse")

20. Andromeda – *al-marʾa (allatī lam tar ba ʿlan)* "the Woman (who Saw no Husband)"

21. Triangulum – *al-muṭallaṭa*, "the Trigon"

The twelve constellations of the zodiac:

1. Aries – *al-ḥamal*, "the Lamb"

2. Taurus – *aṭ-ṭawr*, "the Bull"

3. Gemini – *al-jawzāʾ*, "the Middle One" (originally the name for Orion, "the middle one" probably referring to the three stars in the belt)

4. Cancer – *as-saraṭān*, "the Crab"

5. Leo – *al-asad*, "the Lion"

6. Virgo – *as-sunbula*, "the Ear of Corn" (from the Babylonian name)

7. Libra – *al-mīzān*, "the Scales"

8. Scorpio – *al-ʿaqrab*, "the Scorpion"

9. Sagittarius – *al-qaws*, "the Bow"

10. Capricorn – *al-jady*, "the Goat Kid"

11. Aquarius – *ad-dalw*, "the Bucket"

12. Pisces – *al-ḥūt*, "the Whale"

The fifteen southern constellations:

1. Cetus – *qayṭus* (from the Greek *kētos*); *sabuʿal-baḥr*, "the Predator of the Sea" (in al-Battānī and al-Bīrūnī); *ḥayawān baḥrī*, "Animal of the Sea" (in Isḥāq b. ḥunayn); *dābbat al-baḥr*, "the Brute of the Sea" (in al-ḥajjāj)

2. Orion – *al-jabbār*, "the Mighty One"

3. Eridanus – *an-nahr*, "the River" (from the Greek *potamos*)

4. Lepus – *al-arnab*, "the Hare"

5. Canis major – *al-kalb al-akbar*, "the Greater Dog"; *kalb al-jabbār*, "the Dog of Orion" (in aṣ-Ṣūfī)

6. Canis minor – *al-kalb al-aṣġar*, "the Smaller Dog"

7. Argo (Navis) – *as-safīna*, "the Ship (of the Argonauts)"

8. Hydra – (*al-ḥayyat*) *aš-šujāʿ*, "(the Snake of) the Hero" (Heracles)

9. Crater – *al-kaʾs*, "the Cup" (al-ḥajjāj, al-Battānī, Abū Maʿšar); *al-bāṭiya*, "the Beaker" (Ibn ḥunayn, aṣ-ṣūfī, al-Bīrūnī)

10. Corvus – *al-ġurāb*, "the Crow"

11. Centaurus – *qinṭāwurus* or *qinṭārus* (from the Greek *kentauros*); *al-faras*, "the Horse" (Abū Maʿšar)

12. Lupus – *as-sabu*, "the Beast of Prey"

13. Ara – *al-mijmara*, "the Incense Burner" (from the Greek *thumiatērion*)

14. Corona Australis – *al-iklīl al-janūbī*, "the Southern Crown"

15. Piscis Austrinus – *al-ḥūt al-janūbī*, "the Southern Whale"

These illustrations have been taken from a true to the original facsimile copy[2] of the book prepared circa 1730 in south or central Asia, available at the Library of Congress, USA, which is an exact copy of a manuscript, now lost, prepared c. 1417 for Ulugh Beg. The main parts of the book have been translated into English for the first time by Ishan Hafez in 2013, making the contents of this valuable book available to the English-speaking world.

I am delighted to present Al-Ṣūfī's fine illustrations along with this poem and bring his immortal work to the attention of a new generation of readers.

Abhay K.
Antananarivo, Madagascar

---

2. Fī, ʿAbd Al-Raḥmān Ibn ʿUmar, 903-986. Ṣuwar Al-Kawākib. [1 Muḥarram 820 H 18 February, 1417] Manuscript/Mixed Material. https://www.loc.gov/item/2008401028/.

Awake each night
I savour starlight

(1)

before the blinding sunlight
obliterates you from my sight

(2)

thinking of you, I look at Orion's Belt
and all my sorrows melt

(3)

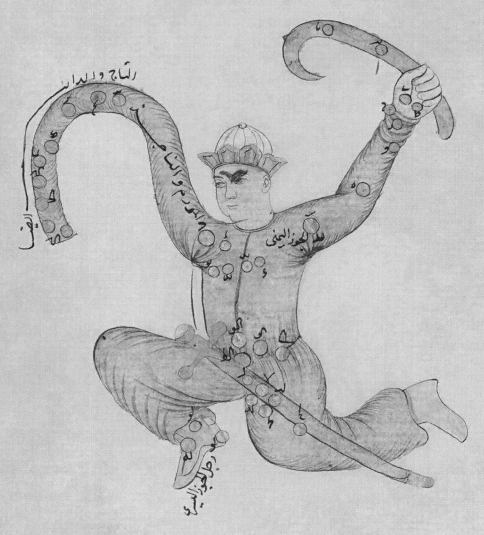

صورة الجبار على ما رى في الكرة

رشاء واليدان

الجوز اليمنى

I seek you in the eye of Taurus
let's ride the bull together, just the two of us

(4)

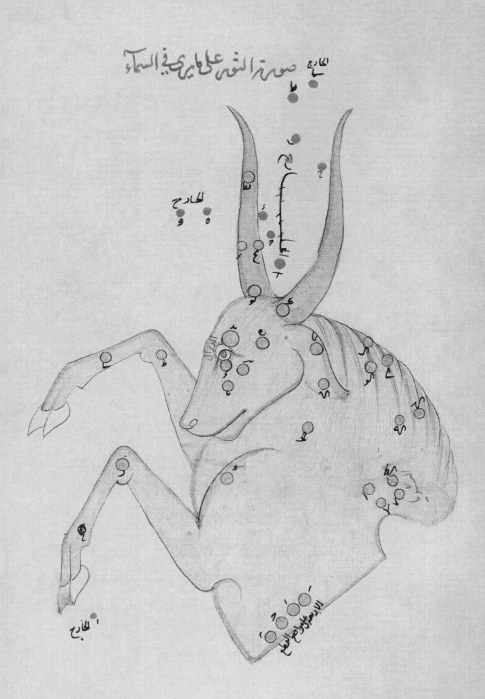

صورة الكلب الاكبر على ما يرى فى السماء

الخارج

I see you in the light of Sirius in Canis Major
shining bright in the night sky, radiant like a laser

(5)

or in the Crux—the celestial lighthouse
reminiscing the days we used to revel and carouse

(6)

where are you my love, my delight
for your one glimpse I chase starlight

(7)

wonder if you are frolicking with Rigil Kentaurus
the nearest star from Earth in Centaurus

(8)

looking for you, I scour Scorpious' Antares
the red supergiant in a mood to tease

(9)

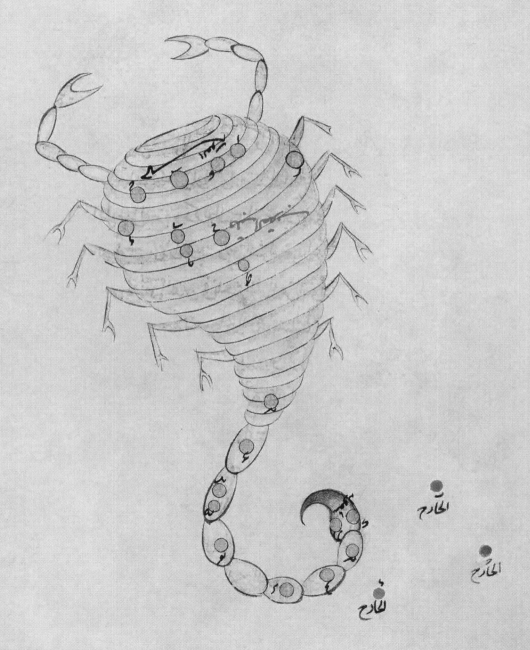

صورة العقرب على ما رى في الكره

I gaze at the wise archer—Sagittarius
how his love arrows have not spared us

(10)

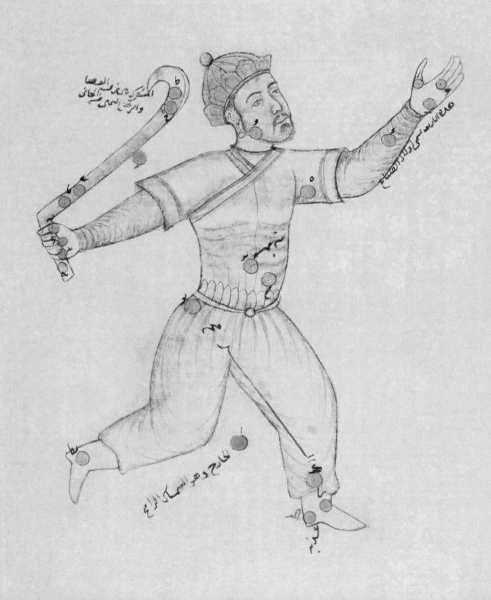

المكبر وبعد ذلك وهو الصفا
والرقص إلى منكبيه

خارج وهو السماك الرامح

my love, come back, and put me at ease
restless, I seek you in bright Arcturus in Bootes

(11)

the ploughman with his two dogs
ploughing the celestial bogs

(12)

I search you in Serpens
your beauty dazzling the heavens

(13)

صورة التنين على ما يرى في السماء
إلكه

setting on a circumpolar sky sail
I closely follow Draco's long tail

(14)

wonder if you're in Pavo—the peacock
dancing on an iridescent cosmic rock

(15)

I feel your aura in Indus
will you and I ever become us

(16)

صورة الفكّتين على ما ترى في الكرة

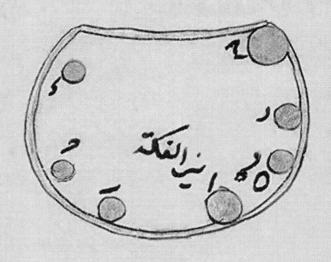

I seek you in the faint Camelopardalis
then in bright Corona Borealis

(17)

have you become Ophiuchus—the serpent bearer
it's time, return to Earth, O celestial wanderer!

(18)

I look fixedly at the mighty Hercules
night after night for your one glimpse

(19)

صورة الجاثي على ركبته على مانرى في الكرة

Are you hiding in the bright Altair in the eagle Aquila
or like the goose Anser in the mouth of the fox Vulpecula

(20)

صورة العقاب على ما يرى في السماء

هذه الستجارة

و الطلس الصغير ين

I wonder if you've turned into Tucana[1]
—the brightly coloured bird of Amazon

(21)

---

1. Tucana is pronounced as Toucan.

or transformed into Grus—the celestial crane
long-necked beauty revelling in meteor rain

(22)

I seek your traces in its blue-white Alnair
shining bright through night, beautiful and fair

(23)

are you hiding in the star Thuban
once the north pole star in Heaven

(24)

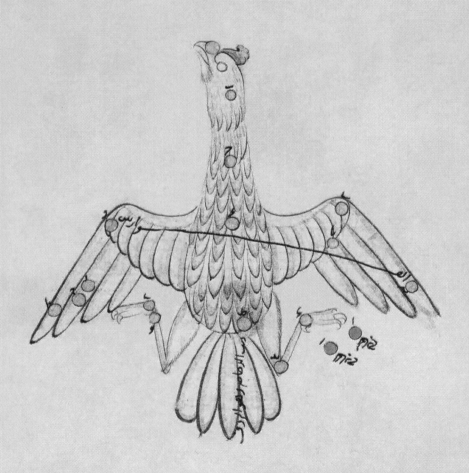

or have you become Cygnus—the swan
to entice me with your charm at dawn

(25)

I imagine you in Pegasus—the winged horse
let us gallop together to the end of the universe

(26)

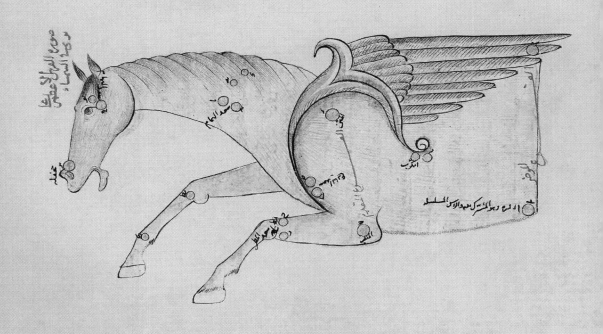

wandering from constellation to constellation
I seek you my sweet love, my one and only passion

(27)

I write poems drinking on the serene
and blessed waters of Hippocrene

(28)

—the spring fountain on Mt. Helicon
hoping my verses will turn you on

(29)

are you hiding in faint Lacerta[2]
away from the eternal cosmic concert

(30)

---

2. Lacerta is pronounced as Lacert.

صورة الشلياق على ما رى فى السماء

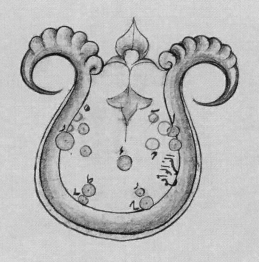

or singing in Lyra's choir
charming all with your mystical lyre

(31)

or visiting the neighbouring galaxy
Andromeda, cruising the cosmic sea

(32)

I turn my gaze towards glittering Cassiopeia
beautiful but vain queen of Aethiopia

(33)

you shine in Alderamin, the crown of Cepheus
—the Aethiopian king—regal and luminous

(34)

صورة قيقاوس على مايرى في الكرة

I look at Hamal—the ram's head in Aries
while he grazes serenely in the celestial prairies

(35)

صورة الحمل على مايرى في السماء

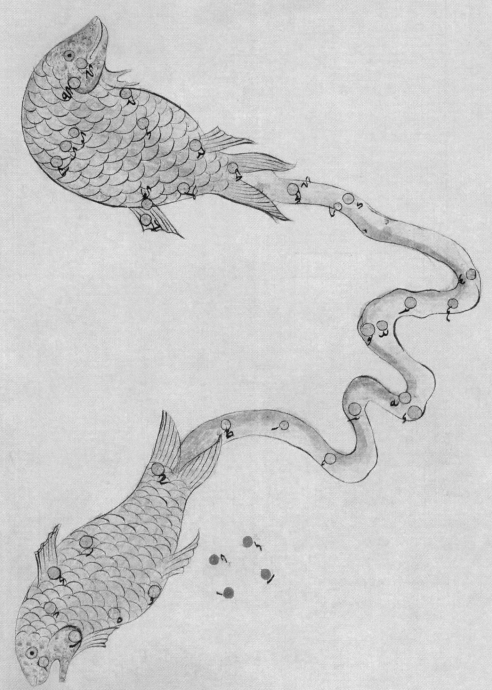

I try tracing you in the faint light of Alrescha in Pisces
love, will my search for you ever cease?

(36)

I soak in the light of Auriga's Capella
hoping to find you in Tringulum's Muthallah

(37)

إسماء

صورة الممسك الاعنة على رىد الكرم

صورة

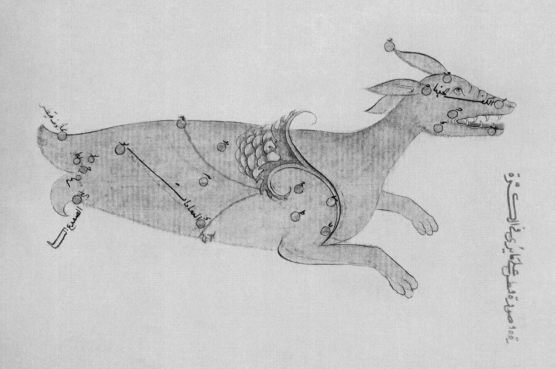

I seek you in Mira and Menkar in Cetus—the whale
being away from you for long I've turned pale

(38)

صورة ساكب الماء
على مايرى في السماء

الصفيح الاول والظليم البضا

I search frantically for you in the pitcher of Aquarius
are you asleep there drunk and delirious

(39)

or have you become the sweet dolphin—Delphinus
inviting me secretly for a tryst, so no one sees us

(40)

صورة الدلفين على ما يرى في الكرة

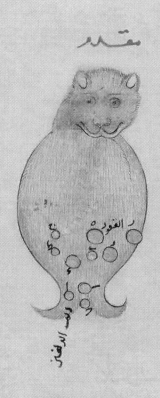

مقدم

القوي

ذنب الدلفين

or are you cantering across the universe, honey
riding Equuleus, the celestial pony

(41)

my heart is pierced with an arrow of light from Sagitta
return to Earth and bring me back la dolce-vita

(42)

صورة المسمكة على ما يرى في الكرة

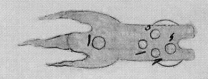

الكرة

صورة السمكة على ما يرى في السماء

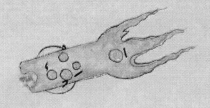

صورة الحوت الجنوبية كما يرى في الكرة

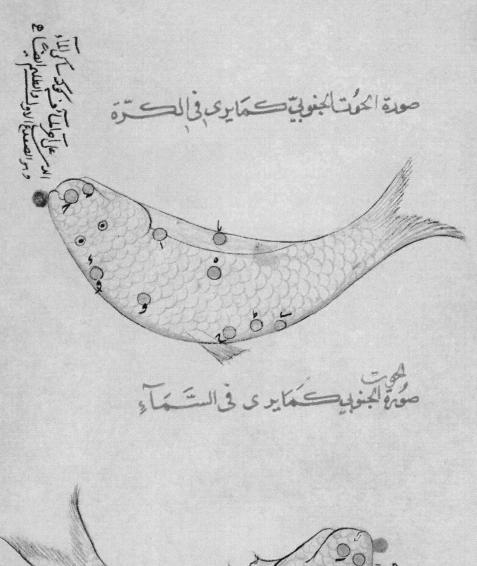

صورة الجنوبي كما يرى في السماء

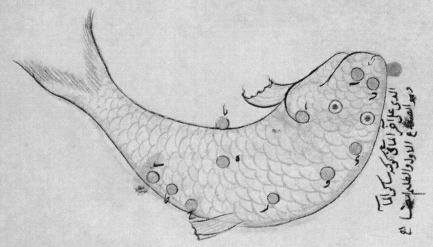

I feel your presence in Piscis Austrinus' Formalhaut
beloved, you are the centre of my world, doubt not

(43)

I croon looking at the neighboring Telescopium
hoping someday we will drink celestial opium

(44)

let's sing together the cosmic anthem
and dance *ananda tandava*[3] in tandem

(45)

---

3. A dance performed by Lord Shiva to express joy.

صورة المجمرة على ما يرى في السماء

I make offerings to Ara—the celestial altar
to help me find you in the teeming star

(46)

I imagine you in Volans—the flying fish
a glimpse of you is my only wish

(47)

I look for you in Octans—the home of southern pole star
let's share a drink at the interstellar bar

(48)

wonder if you've turned into a Chamaeleon
camouflaging yourself as a seductive alien

(49)

or have you become Apus—the bird-of-paradise
soaring in the southern skies

(50)

I glance at Algol, the demon star in Perseus
wondering if it has cast its evil spell on us

(51)

I look expectantly at Dorado—the dolphin fish
being with you, my love, is all I wish

(52)

are you sitting at Mensa's cosmic table
looking towards Earth, reading a cosmic parable

(53)

or gambolling with Hydrus—the little water snake
beloved! return to Earth for my sake

(54)

awake through the night till morning at six
for you, I scour bright Ankaa in Phoenix

(55)

I seek help of the sea goat Capricornus
hoping he can unite us

(56)

I look through Sculptor in Cartwheel Galaxy
the thought of finding you there fills me with ecstasy

(57)

my eyes meander towards Sagittarius A*
seeking you in the heart of the Milky Way

(58)

certain, someday we'll meet up there
sooner or later, I need not despair

(59)

you glimmer in the diamond-shaped Scutum
its shield protecting your cosmic kingdom

(60)

صورة الميزان على عاري في الكرة

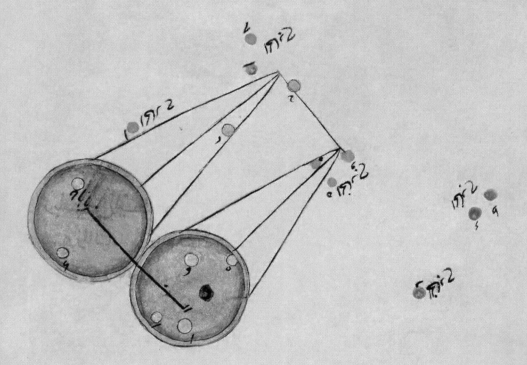

you may be busy balancing Libra's scales
while I try to trace your heavenly trails

(61)

have you become Virgo la belle
peeping shyly through your celestial veil

(62)

I espy you in the solitary Alphard
giant sea snake Hydra's vanguard

(63)

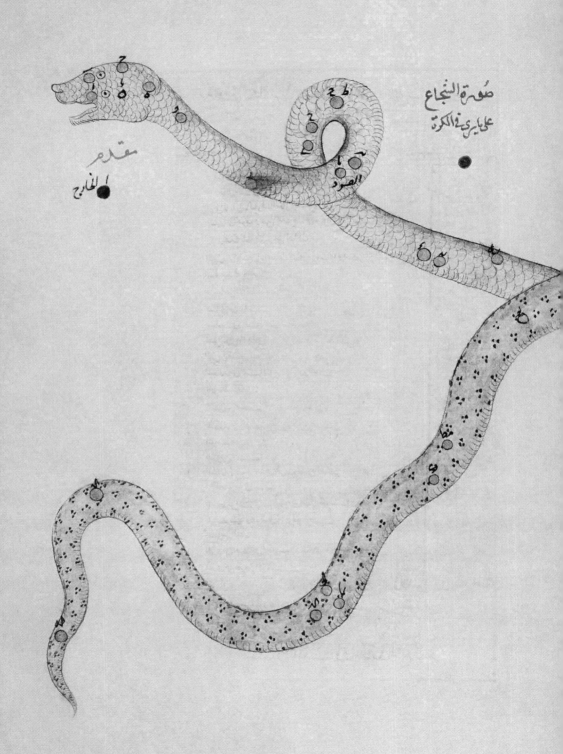

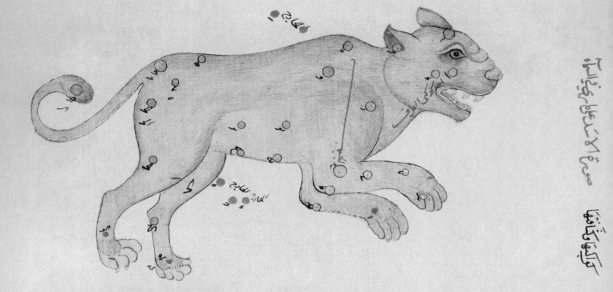

صورة الاسد كما يرى في الكرة

الاسد بصورته الحقيقية

I peer into Leo's heart—the bright Regulus
hoping someday the universe will bless us

(64)

let's drink the sweet wine from the cup of Crater
whenever, wherever we meet, sooner or later

(65)

here on Earth or in some distant star constellation
to you, always and forever, I vow my love and devotion

(66)

I focus my eyes on dim Sextans
seeking you in the cosmic gardens

(67)

I sense your presence in bright Gienah in Corvus
let raven be the messenger between the two of us

(68)

صورة الغراب على ما يرى في الكرة

جناح الغراب

المنقار

your golden hair burnishing like Berenices'
in night's solitude I crave for your soft kisses

(69)

are you sailing in the cosmic sea with Vela[4]
take me along my love to the celestial *mela*

(70)

---

4. The modern constellations of Vela, Carina and Puppis were
earlier part of the original constellation of Argo Navis.

where stars dazzle with great lustrous charms
let's ride the ferris wheel with galactic arms

(71)

I see you blossoming in Carina's bright Canopus
pleasing as a freshly bloomed crocus

(72)

I wonder if you're sitting on the Puppis' deck
waiting to celebrate our union with a cosmic cake

(73)

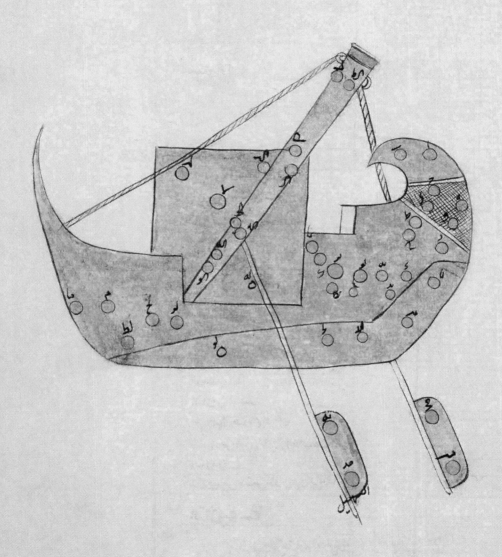

painting our portrait on Pictor's easel
or waiting to tango in cosmic drizzle

(74)

are you effusing your fragrance from Antila[5]
you smell like the Madagascan vanilla

(75)

---

5. Antila means pump.

I bask in Canis Minor's bright Procyon
remembering you, ah! those days of halcyon

(76)

صورة الكلب الاصغر على مايرى فى الكرّة

المرزم

الشعرى الشامية

I fantasize you in Columba's Phact—the ring dove
awake whole night, pining for you my love

(77)

perhaps you're chiseling a poem with Caelum[6]
or making a net of crosshairs in faint Reticulum[7]

(78)

---

6. Caelum means chisel.
7. Reticulum means small net.

thinking of you voyaging the universe
I paint your portrait in verse

(79)

my mind sways like Horologium—the pendulum clock
pacing back and forth, I wait for you at the cosmic dock

(80)

I turn to Castor and Pollux in Gemini
to guard you from the evil eye

(81)

صورة التوأمين على ما يرى في الكرة

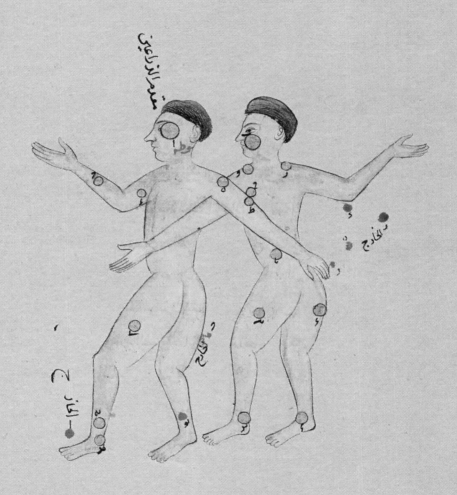

burning with desire like Fornax's furnace
I'm anxious for a glimpse of your charming face

(82)

I canter off my gaze towards Monoceros—the unicorn
perhaps you are riding it holding its horn

(83)

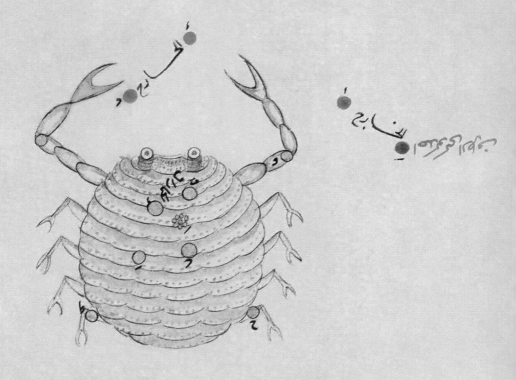

I fathom you in Cancer's bright Altarf
standing at the edge of the cosmic wharf

(84)

have you become the gorgeous Lynx
ambling in the skyways without any kinks

(85)

I gape at the little lion in faint Leo Minor
hoping to dine together at a fine cosmic diner

(86)

السماء

صورة الدبا لاكبر على مايرى فى الكرة

سوخر

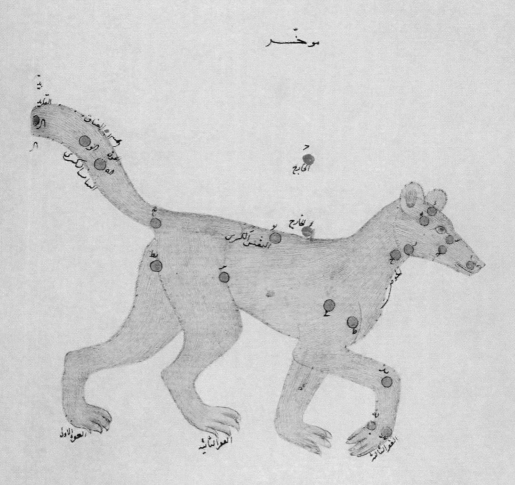

I wonder if you glitter in Dubhe in Ursa Major
like diamonds, my celestial treasure

(87)

or shine like Polaris—the pole star in Ursa Minor
may its light guide me gently to wherever you're

(88)

صورة الدب الاصغر على ما يرى في السماء

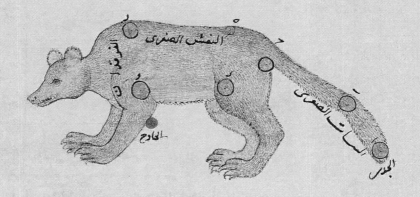

النقش الصغرى

الذنب

الجدح

اسنا الصغرى

الجدي

I see the celestial wolf Lupus gazing from the skies
pining, I look skyward, tears welling up in my eyes

(89)

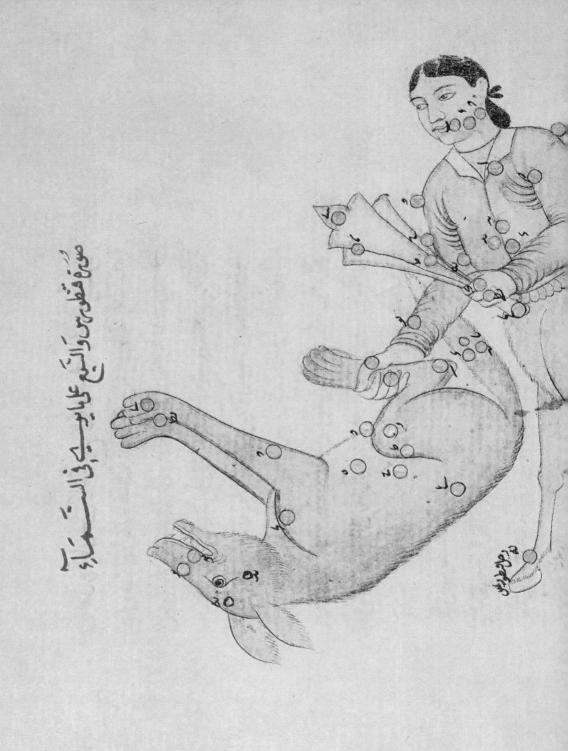

I inspect Norma, the carpenter's square
wondering if you're hiding there somewhere

(90)

I seek help from the compass Circinus
to find you my sweet Venus

(91)

الكرة

صُورَة الاكليل الجنُوبي على مايرى في السَّمَاء

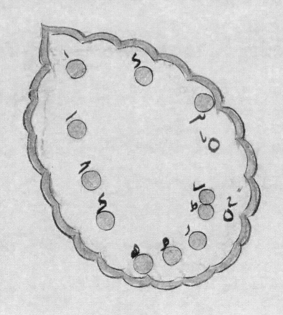

I seek you in faint Corona Australis
craving for our union—the eternal bliss

(92)

are you in bright Atria in Triangulum Australe
wait for me there, I'm setting sail

(93)

I gaze at the bright Cor Caroli in Canes Venatici
are you walking the dogs in the Whirlpool galaxy

(94)

I shift my gaze at the minor Microscopium
wondering you're there and keeping mum

(95)

I look out for you in Musca—the celestial fly
my beautiful maiden, for you I scour the sky

(96)

I examine the bright Arnab and Nihal in Lepus
perhaps you're leaping like a hare in Christmas

(97)

صورة الأرنب على ما يرى في الكُرة

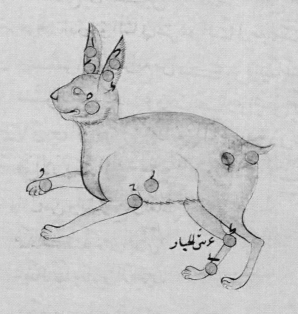

عين الجبار

صورة الأرنب على ما يرى في السَّمَاء

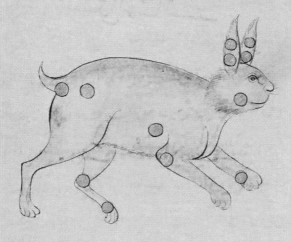

come back to Earth or take me along
I live for you, you're the only one I long

(98)

at last I ask Pyxis—the Mariner's compass
to show me the way, I'm getting anxious

(99)

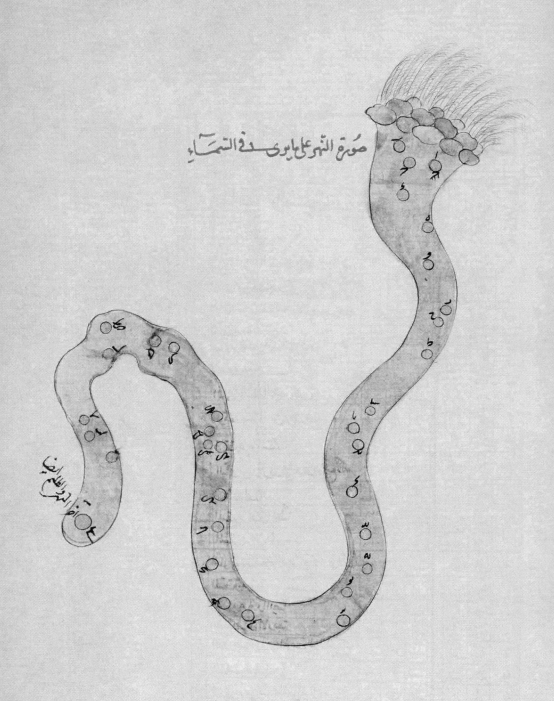

ah! I find you in Achernar at the end of Eridanus
the benevolent universe finally unites us.

(100)

Photo by Alina Medvedeva

Abhay K. is the author of a dozen poetry collections, including *Stray Poems* (Poetrywala), *Monsoon* (Sahitya Akademi, 2022), *The Magic of Madagascar* (L'Harmattan Paris, 2021), *The Alphabets of Latin America* (Bloomsbury India, 2020), and the editor of *The Book of Bihari Literature* (HarperCollins, 2022), *The Bloomsbury Anthology of Great Indian Poems*, *CAPITALS*, *New Brazilian Poems* and *The Bloomsbury Book of Great Indian Love Poems*. His poems have appeared in over 100 literary magazines, including *Poetry Salzburg Review*, *Asia Literary Review* among others. His 'Earth Anthem' has been translated into over 150 languages. He received the SAARC Literary Award 2013 and was invited to record his poems at the Library of Congress, Washington DC in 2018. His translations from the Sanskrit of Kalidasa's *Meghaduta* (Bloomsbury India, 2021) and *Ritusamhara* (Bloomsbury India, 2021) have won KLF Poetry Book of the Year Award 2020–21. www.abhayk.com